HOW TO MAKE YOUR OWN PROFESSIONAL LOCK TOOLS, VOLUME 2

—By: Eddie The Wire

Loompanics Unlimited Port Townsend, WA

Other books by Eddie the Wire:

The Complete Guide To Lock Picking
How To Bury Your Goods
How To Make Your Own Professional Lock Tools, Volume One
How To Make Your Own Professional Lock Tools, Volume Three
How To Make Your Own Professional Lock Tools, Volume Four

HOW TO MAKE YOUR OWN PROFESSIONAL LOCK TOOLS, VOLUME 2

Printed in USA

Published by:
Loompanics Unlimited
PO Box 1197
Port Townsend, WA 98368
Loompanics Unlimited is a division of Loompanics Enterprises, Inc.

Cover by Michael Hoy
Illustrations by Kevin Martin

ISBN 1-55950-048-4

CONTENTS

A WARNING NOTE FROM THE AUTHOR

As always, if you attempt to manufacture or use any of the equipment described in this book, Eddie The Wire and Loompanics Unlimited deny any and all responsibility for errors, malfunctions, or problems arising from your acts. *This book is sold for informational purposes only.* Remember: the mere insertion of a lock tool in somebody else's keyway without their permission is a crime. Strengthen yourself with knowledge, but do not embark on a life of crime. *NOTE: The material in this book cannot be effectively understood unless you have read* HOW TO MAKE YOUR OWN PROFESSIONAL LOCK TOOLS, Volume One *thoroughly.*

-Eddie The Wire

CHAPTER 1
HOW TO "STONE" A PICK SET

The common method of producing commercial lockpicks today is to cut or shear the whole profile from a strip of steel using a punch press and a male-female die. This process leaves grooves and striations on the sides of the stamped piece (the edges) due to variations in the dies and wear and tear. Often, these rough edges prevent the pick from working smoothly in the lock, and may even hang up on a tumbler momentarily, causing the tumbler to move away from a shear line. This can be very frustrating to the would-be lock specialist, especially if he is trying a raking-type procedure.

Custom picks that have been correctly rough ground and file/drawfile finished according to my procedures (see HOW TO MAKE YOUR OWN PROFESSIONAL LOCK TOOLS, VOLUME ONE) have few if any rough edges and may be used stock. For the perfectionist, though, and for the person who uses commercial picks, a procedure called stoning is very helpful. The same sort of flat bench stone that is used for sharpening knives is employed for stoning. This may be natural or synthetic (made in a furnace) and a small slip or gouge stone (which is curved to sharpen curved edges) is also helpful.

I use a synthetic stone with a rough and a fine face. Before stoning a pick set, make sure that the stone will not move around on the workbench. One way you can do this is to clamp or nail four boards around the stone. A better answer is to take a piece of plywood not as thick as the stone and trace the outline of the stone on the plywood's surface. Then cut out around the line trying to get a very tight fit. If the fit of the stone is too loose, a shim can be glued to the inside edge of the plywood. This stone holder may be clamped or nailed to the work-bench, close to the edge.

Another prerequisite for stoning is some kind of oil. Light machine oil (the three-in-one kind), heavy motor oil that has been cut or diluted a little with solvent, or some kind of linseed oil is good. Some people use a liberal amount of WD-40. The object of the oil is to float away the particles of steel that would otherwise become imbedded in the stone's surface and impair cutting efficiency, so keep the stone well oiled. A new stone may even have to be soaked in oil to prevent its surface from becoming too dry initially. For regular stoning, about two teaspoons of oil or so is good to start.

The actual stoning itself is rather simple. Inspect the edges of the pick and look for uneven or wavy marks or any surface imperfection. Shining a light at a slight angle to the pick's edge (almost parallel to the edge) will reveal many imperfections that the overhead light washes out.

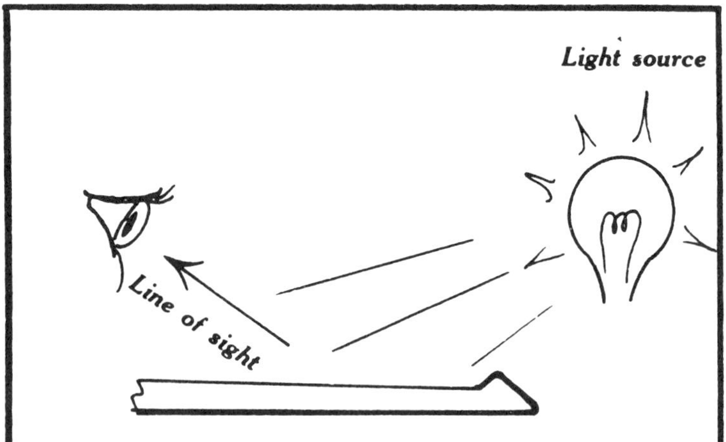

Figure 1.

This lighting procedure is also useful in the skill of "impressioning", which will be covered later on. Once the needed edges are located, stone them smooth by working the edge squarely along the stone's surface in a direction of travel that is parallel to the normal line of travel for the pick when it is worked in the keyway of the lock. This produces only parallel scratches that cannot hang up the pick even minutely.

Don't apply excessive pressure, just wear away the

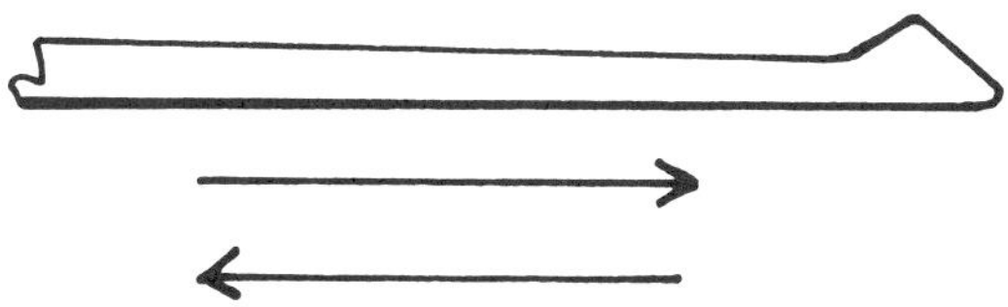

Stone the pick this direction only

Figure 2.

roughness slowly and evenly. From time to time wipe the pick free of "floating oil" and recheck the edge for smoothness. *Do not* run your fingers along the edge to check, because if you do the procedure correctly you will have tiny, long splinters of steel on the pick's surface that are impossible to remove from a finger, so be careful. Try also to avoid excessive stoning past the point of smoothness, because you can change the dimensional contour of the pick profile. Be especially careful not to change any pick angles or contours. To get into those small spaces it may be necessary to use the slip or gouge stone carefully. Another good technique here is to purposely round one of the edges on your bench stone with a hard steel implement like a file tang or a drill bit, and use the rounded edge of the stone to smooth inside curves on the pick.

Once the pick is rough stoned on all edges, if you have a fine stone take a few passes to complete the job. It doesn't take much for the fine stone. Normally only the edges of the commercial pick will require stoning, although I have seen a few with mill scratches still on the faces or sides of the pick. These should also be removed, if present. If you are stoning your custom pick set, go directly to the edges that have been previously ground and omit the lower edge because the regular profile grinding procedure does not touch this edge.

The faces on a custom pick are usually okay, but check them also. Sometimes I will stone the faces of raking type picks, like diamonds and one and two tumbler profile picks. This provides a rougher surface that will not contact the keyway sides but does hold a lot of lubricant on its surface, which is often beneficial for these techniques.

CHAPTER 2
BLUING (OR RE-BLUING) YOUR PICKS

Once the stoning procedure is done, it is wise to blue or re-blue the picks. Especially important are picks that will be carried taped to the skin, or in porous carrying pouches for any length of time. If they are not blued they will rust for sure, and the nice smooth edges will re-roughen and the steel will stain.

Prior to bluing the pick, it must be de-greased. The easiest way to do this is by washing/scrubbing in a solvent like naptha (lighter fluid) held in a container. Once all traces of grease and oil are removed, (also no finger prints, OK) immediately start the bluing process according to the manufacturer's directions.

I prefer one of the cold-process wipe-on bluing kits that are available for gun touch-up. If you are immersing the pick, fine. If you are brushing on the bluing agent, try to avoid streaks or overlaps. Also, do not leave the pick in the solution for too long, and remember to thoroughly rinse in water afterwards. Once the pick has been rinsed it is very susceptible to finger prints that will etch themselves in -- a bad thing to do to a lockpick.

For the final touch, after the pick is dry, apply a small amount of lubrication, either WD-40, machine oil, or even one of the new synthetic teflon-derivitive oils. This provides pick surface protection and smooth operation. Make sure that the lubricant used does not stain or affect your carrying pouch material.

CHAPTER 3
HOW TO GET TOP—QUALITY STEEL IN THE PROPER DIMENSIONS

While we are on the subject of picks, let me tell you how to get a really top quality steel in the proper dimensions. Simply go down to the hardware or super store and get a plumbing snake (the do-it-yourself kind). If more than one width of snake is available, get the thinnest size. Notice that the snake is nothing more than a tempered steel strip with a funny looking end and a handle. Picks cut from this material are very good and the cost is not great. Also, the pick stock is less wide than the traditional automobile shim stock (½″ wide), so less cutting is involved. You may have to build up a bigger handle, though. For the very finest in steel however, continue to use the auto shim stock.

CHAPTER 4
DRILLING THE STOCK

Another topic that deserves treatment in this grab-bag is how to drill the auto shim stock. Several concealment applications (in pens, jack knives, on strings) demand that the pick have a hole drilled in it. Anyone who has tried, however, knows that the steel is usually tougher than the drill. The problem is that the point of the drill cannot cut enough of a dent in the pick's surface to allow the large twin cutting flutes of the drill to function.

The answer to that is simple -- just use a large, sturdy prick-punch to produce the start of a hole or depression. The punch should be very sharp, or the depression sides will not touch the drill. Remember to strike it sharply. Once the hole is started like this, it should drill easily.

For large holes it is better to drill a smaller hole first then follow up with the larger one. Using lower speed for the bigger drill also helps to eliminate "chatter". If you can drill five or six

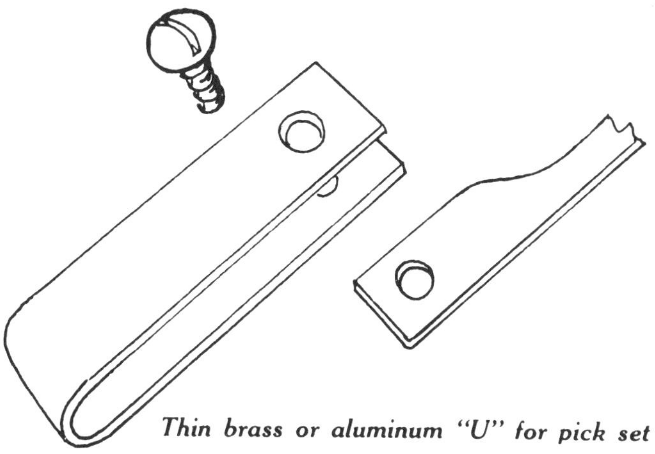

Figure 3.

picks, it is easy to bolt them all together, then grind away any differences among them in contour (not working end, but handles), and then fashion a case for them out of sheet brass or aluminum to make a pocket pick set.

Some automotive feeler gauge sets are provided with a removable bolt that allows access to individual feeler blades. A set like this can easily be altered to a pick set, even using some of the blades as pick stock.

CHAPTER 5
PRODUCING LARGE QUANTITIES OF CUSTOM PICKS

Custom picks are by far the best tool obtainable for lock work, but they can be a hassle to produce in large quantities. Consider that the working set should have at minumum two heights of diamond and two heights of lifter, with three or four widths and thicknesses of tension wrench and snake picks as well. Now think that the average lock specialist must carry a personal set, have an emergency hide-out set, and also a large set in his tool box.

Let's look at some of the techniques involved in producing large quantities of picks. The best way, of course, is to stamp the profile -- like commercial picks are produced. If you have access to machine shop facilities, a die that produces rough profile blanks is worth its weight in gold. You could engineer a die to cut down stock to the minimum necessary and grind the finished profile.

Failing this (and most people will) the next method is to Mass-mark and finish the picks. Start by grinding one rough profile blank (see HOW TO MAKE YOUR OWN PROFESSIONAL LOCK TOOLS, VOLUME ONE for an explanation of this term), or better yet, cutting it from a sheet of heavy plastic, say one sixteenth of an inch thick.

Auto shim stock can be sprayed with diemaker's blue or toolmaker's ink to produce a markable surface. These spray can marking products can be applied to large amounts of stock simultaneously, and are available at any industrial supply outlet.

After the ink is dry, the patterns can be individually traced on the blanks with a metal point or scribe. When tracing, remember to keep the point turned in towards the pattern and not away. If the side of the scriber rides on the patern edge, but the actual point of the scriber is away from the pattern edge (which is what happens when you tilt the scriber) the pattern will be incorrect.

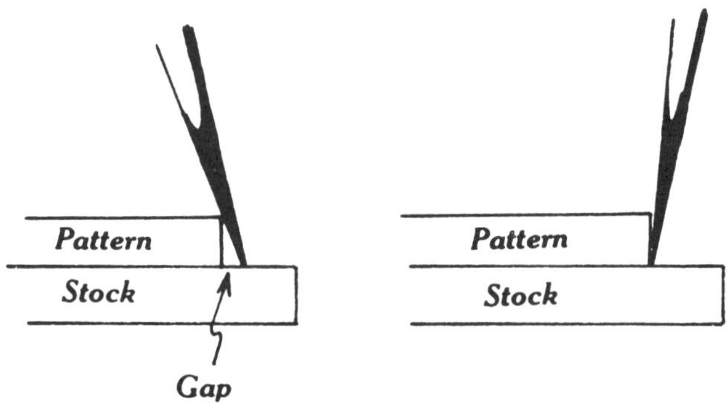

Incorrect scriber angle *Correct scriber angle.*

Figure 4.

The rough profile tracing concept can be further applied to actual finished profiles. Remember, the rough blank can be rescribed to any final form, it is flexible. The blank that is scribed to a particular profile must be cut to that profile, unless you want to clean off the dye or ink and start over. The pattern you use for each specific profile should be the exact size you want -- perhaps tracing directly from a pick you use a lot is a good idea.

CHAPTER 6
USING PAPER PATTERNS

At this point let me warn you about paper patterns. The technique of photocopying picks and gluing the resultant profiles onto blanks directly to use as a pattern is a great idea except for the "quenching-in-water" step necessary with all pick grinding. When the pattern goes through ten or twelve quenches, it comes up soggy and blurry. Speaking from experience, I advise either waterproofing the paper or using toolmaker's ink and a scribe.

CHAPTER 7
GRINDING ROUGH PROFILE BLANKS

Now that you have hundreds of blanks with profiles laid out on each one ready to grind, is there an easy and fast way to rough grind that profile? Yes -- the answer is to push harder against the grinding wheel (not so hard that you load down the wheel though -- listen to the pitch of the motor and slack off pressure if this pitch drops a lot) and plan your cuts in full widths. The problem with grinding faster is that the heat also builds up much faster. The problem then becomes one of eliminating any delay in the quenching process and get back to the wheel quickly. One easy answer to this is to position the quenching water dish no more than six inches from the wheel face, and directly below the wheel. With this arrangement you can easily flick the pick down with one hand and swirl it in the water, then zip it quickly back to the grinding wheel.

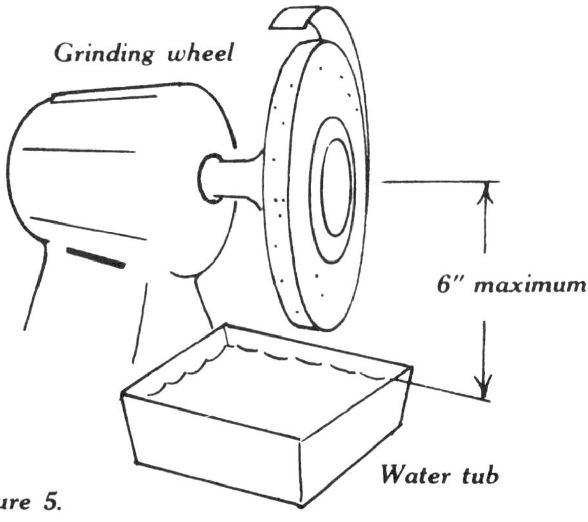

Figure 5.

In HOW TO MAKE YOUR OWN PROFESSIONAL LOCK TOOLS, VOLUME ONE, I talk about counting to three, then quenching the pick, then back to the wheel for a three count. This concept of rhythm should be expanded to produce good grinding speed. The entire sequence could sound like this: *1-2-3-dip-swirl-zip-1-2-3-dip-swirl-zip*, and so on. The less time wasted unnecessarily, the faster the cutting will go. Try it. It is surprising how quickly a pick can be cut if you work up a good rhythm.

I also mentioned taking full width cuts. This means that you should start the edge of the cut at the edge of the wheel, and after producing a cut perhaps a thirty-second of an inch deep, shift over exactly one wheel width and start the next cut.

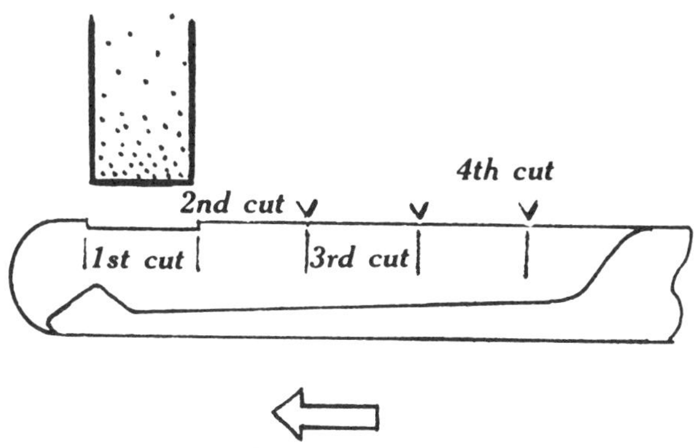

Figure 6.

Of course, unless the pick profile is exactly two or three wheel widths wide you will have a smaller cut at the end, but try to get maximum usage from each pass of the wheel.

Another thing to try depends on your grinding technique. If you usually grind with the thickness of the pick held very close to the axis of wheel rotation, then you may want to try multiple blank cutting. Any number of blanks up to five can usually be cut as one unit simply by squaring up a stack of blanks and securing them together.

Some common ways of securing include taping, clamping and gluing. The taping (vinyl electrician's tape) is the easiest, and the tape can be cut and secured in different places as the grinding progresses.

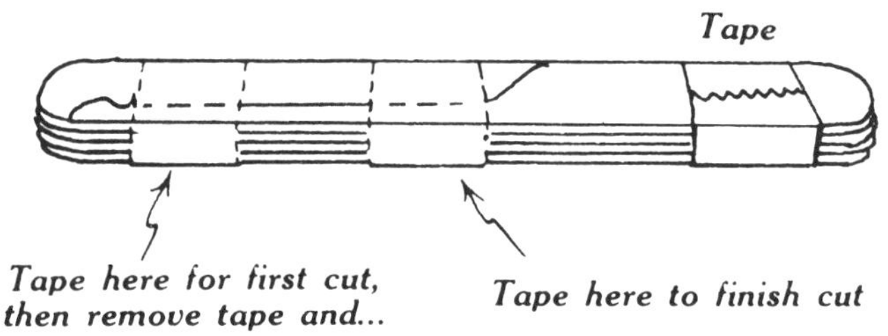

Figure 7.

The clamping must be done with care, and the pack re-squared carefully if the clamp must be loosened and reapplied, but the handle the clamp affords is nice. (I use vice-grips.) Remember to rely on your sense of touch when grinding as far as heat build-up, or you will ruin picks. Fingers always burn before picks do.

Plastic laminate type contact cement provides the best gluing available, and the stack can be soaked in solvent to release the individual picks after file finishing.

In general, the larger diameter grinding wheels can handle a larger number of blanks in the stack, but start out with only two blanks, and check them both carefully for accuracy after the initial rough grind. If they are both true, increase the number in the stack by one. Eventually you will find out what number of blanks can be accurately ground with your individual technique.

Once the stack has been rough ground, do the file finishing on the blanks while still stacked because the added thickness helps to keep the file square to the pick surface. At the very least, this technique should be tried to produce rough profile blanks, and also simple diamond profile picks.

Now that the picks are rough ground and file finished, you must remove all traces of toolmakers ink with solvent, and then stone the pick if you wish, but remember to blue it.

The final step would be applying a handle, and I prefer the vinyl tubing slip-on for a set of production line picks. Fifteen or twenty handles can be heated in an oven and applied when soft and flexible, they will soon shrink to fit.

CHAPTER 8
PRODUCTION LINE TECHNIQUES FOR TENSION WRENCHES

Tension wrenches are also liable to production line techniques, especially the music wire type with two flats ground on the working end. The first step is to cut six or seven to length plus one inch. The ends should not be ground flat at this point. Following cutting, grab the last quarter-inch of wire stock with a pair of vise-grips and lock then. You may even wish to wire them closed.

Now position a flat table surface at least four inches square so that the wire, when the vise-grips are laid on the table, hits the grinding wheel face exactly even with the axis of the wheel.

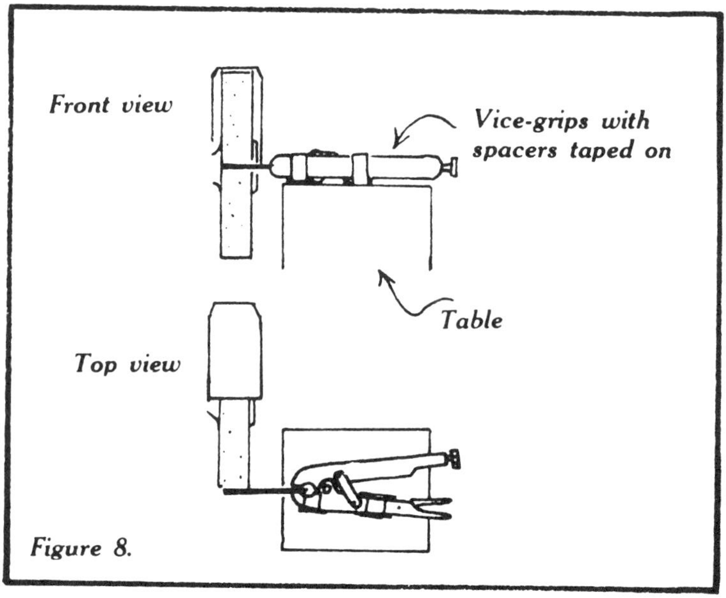

Figure 8.

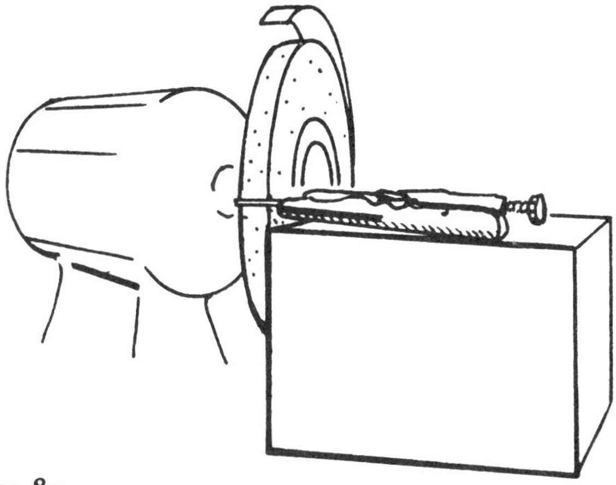

Figure 8a.

Once this is accomplished (you may have to improvise a couple of clamps, but it is worth it), grind the last inch of the other end of wire stock with a flat on it of the desired width. When that is done, don't unclamp the vise-grips but just flip them over so that their other side rests on the table surface, and grind the other flat on the wire end. If your vise-grips don't lay flat initially or wobble, you can find the low spot on the tool and pad it with tape and spacers -- just so the vise-grips can be flipped 180 degrees.

Once the two flats have been ground to equal thickness or width, remove the wire stock from the tool. Commercially available bending jigs will allow you to bend the working end 90 degrees, or you can improvise a jig by sawing a slot in a piece of angle iron and inserting the end of the wire.

An even better method is to drill a hole in a sandwich consisting of angle iron with a two-by-four backing.

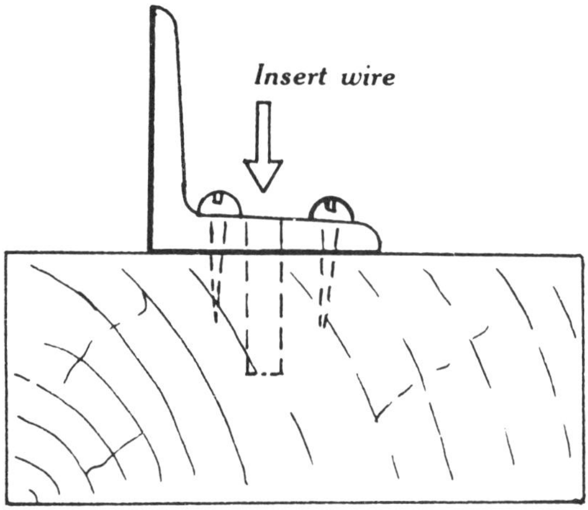

Figure 9.

The hole should be three or four thousanths of an inch larger in diameter than the wire stock, and as deep into the two-by-four as you want to go before making the bend. With this method, the bend always comes at the same point on the wrench -- just insert, bottom, and bend. With the wire bent, it is an easy matter to cut to length and grind one or both ends smooth of cut marks.

Double-ended tools can be made with the same technique, and slanting the flat faces 90 degrees from each other can easily be done by eye while the tool is in the vice-grips prior to being clamped.

That covers methods for tension wrenches; most of the locksmiths and other people that I have interviewed prefer this style of wrench. One such locksmith is using one now professionally, in preference to commercial models. Of course, remember that rough profile blanks can be cut down to become tension wrenches as well (see HOW TO MAKE YOUR OWN PROFESSIONAL LOCK TOOLS, VOLUME ONE).

CHAPTER 9
THE SLIM-JIM

The next subject to cover is the slim-jim. Instructions (brief) on manufacture and use of a slim-jim were given in HOW TO MAKE YOUR OWN PROFESSIONAL LOCK TOOLS, VOLUME ONE but a new and better model is currently in use. In fact, a recent publication lists six versions of this tool, plus a Corvette-opening tool. My tool is cut from an aluminum shop ruler, available at most hardware stores. The ruler I use is three feet long and one and a half inches wide, but there are many sizes on the market. From this ruler, cut a twenty-three inch long piece with a hacksaw, and grind the end as shown in figure 10.

The measurements are given so that you can spray with toolmaker's ink and scribe the pattern on the ruler directly. With the right lighting, you can even omit the toolmaker's ink. These rulers are all tempered aluminum, so watch for heat build-up as usual, and quench frequently. If your first stop is an art store, look for a flexible tempered steel ruler in a length that is close to twenty-three inches, and use it instead.

Now that the end has been cut to profile, the thickness of the working end must be specially sharpened to avoid catching on the weather-stripping on all car windows.

Figure 11 shows the profile for sharpening. It allows two-way tool insertion and no hang-ups. A hole should also be drilled in the handle end of the tool, one-eighth inch in diameter and located a quarter of an inch from the edge. Finally, get a large heavy rubber band and double it around the shank of the tool. Working the tool is fairly simple. Simply insert the working end between the weather stripping and the glass at the driver's or passenger's side, straight down from where the lock cylinder is. Disregard where the inside button is. The object is

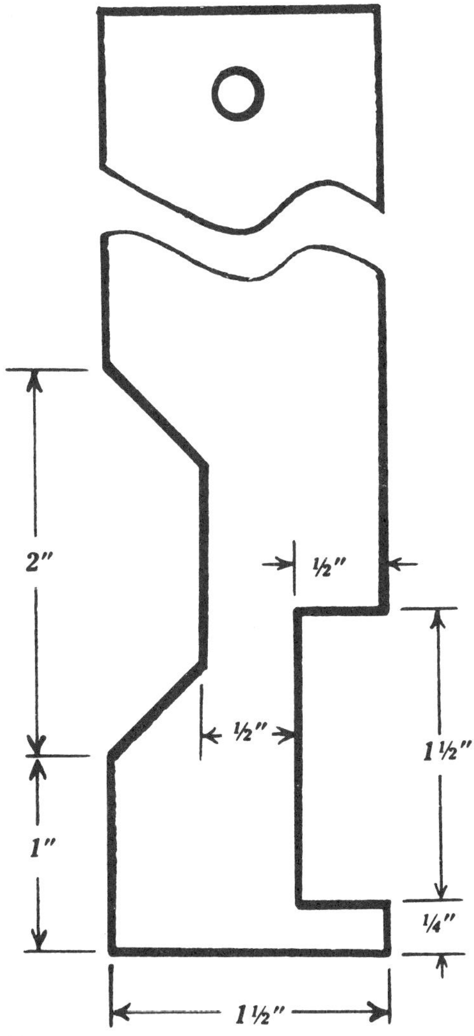

Figure 10.

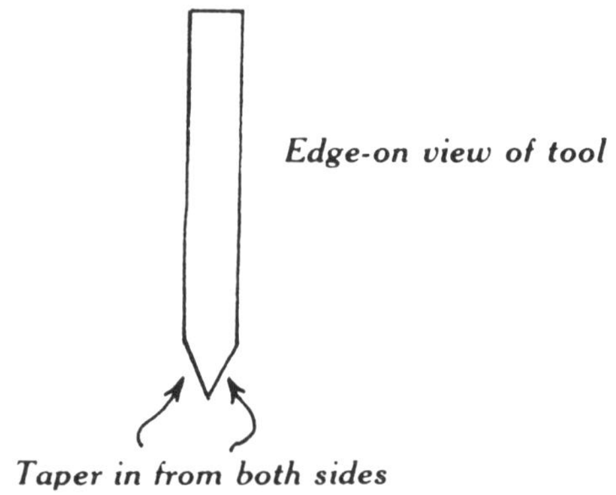

Figure 11.

to hook either the camming mechanism or linkage that is attached to the back of the lock cylinder, or hook the rocker shaped latch assembly located at the back edge of the door. Once hooked, a simple pull, or alternately, a push down, will unlock the door.

There are finer points, though. One is to lubricate the weather stripping with a spray silicone or hand cream for easier insertion. Some cars, notably FORDS, have two layers of stripping -- in this case, go through the middle. A fifteen to twenty five degree bend in the tool end may be very helpful in contacting the lock mechanism -- once you get it right, keep it bent. Occasionally, while working the tool, the door unlocks but does not open so try the button or the handle once or twice while working, and watch the inside button.

Another point is that smaller cars, especially of foreign manufacture, will have bolts that obstruct the straight down maneuver, but an angling movement will avoid these. Some cars even open with a sideways push on the tool shank.

Now to the rubber band. Before insertion in the door hold the tool with the hook on the cylinder and slide the rubber band until it is level with the top of the door. This will provide an accurate measure of how far down to insert the tool. In fact, if the inch markings are intact on the ruler, you can go by those and save the extra time involved in rubber banding. After all, time is of the essence while opening.

I strongly advise that you keep records of each attempted opening so that you can duplicate it next time. Some people keep a notebook listing make, model and comments.

Some makes and models of cars from 1979 on have metal shields or framing members that impede this slim-jim. For those tricky situations, I recommend a length of three thirty-seconds of an inch diameter music wire with a handle looped in one end and a right angle bend on the other end.

The right-angle member should be an inch and a quarter long, and the whole tool should be twenty three inches. This tool is inserted by putting the end of the bend straight down through the weather stripping (a little sharperning may be helpful here in avoiding metal retainer clips -- but not too sharp) then bringing the shank of the tool up ninety degrees to normal working position. This puts the right angle bend inside the door with a minimum of forcing. From there the tool can be used as normal, although it will have a tendency to hang up easily. When that occurs, don't panic, just work the tool back and forth slowly and it will free up.

A rubber band on the shank of this tool is very helpful in locating the locking mechanism depth properly. The hole on the other end can be used as a lasso to open mushroom style inside knobs, the threaded wire that remains when the knob is removed, and some types of anti-theft smooth knobs. Just bend

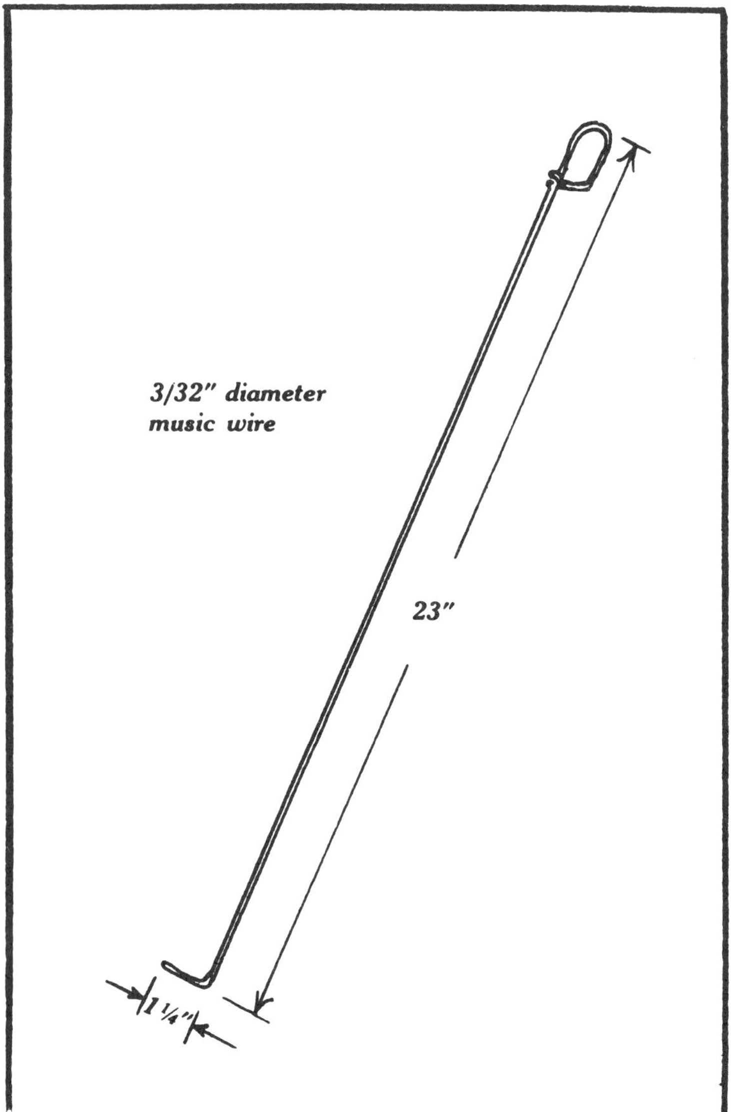

Figure 12.

at an angle and insert, as the tool is pulled over the knob it will cant and dig into the surface, pulling up the knob.

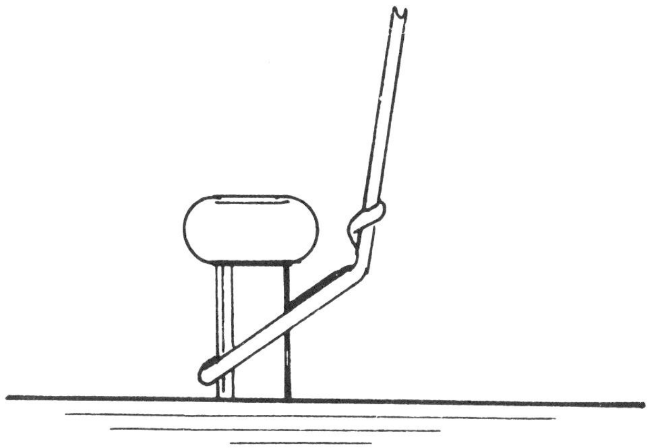

Figure 13.

Another well-known (to most people I know, anyway) dodge is tying some fishing line to the hole, then leading the end back through, up the tool shank, and out the car. The noose that results can be looped over a button and tightened by a pull on the line, then the whole tool pulled to lift the button. Oh, by the way, avoid all of the other car opening tools on the market, most of them are designed to operate the inside button, and only the slim-jim is really reliable.

CHAPTER 10
CARRYING POUCHES, STASHES, CACHES, AND SO ON...

Let's talk about carrying pouches, stashes, caches, and so on next. In HOW TO MAKE YOUR OWN PRO—FESSIONAL LOCK TOOLS , VOLUME ONE I gave a few ideas on cases like wallets and credit card cases, but a *really* professional case is easy to manufacture. I recommend that every lock specialist carry a set of tools on his person -- it's terrible to be locked out and know you could get in if you had the tools.

One of the easiest carrys available is to iron on an inside pocket. Jeans or heavy cloth clothes take this best of all. Many stores sell iron-on repair patches. Just get a package that has pieces of the right weight of fabric, the right color if it will show outside, and of course, big enough to accomodate two or three picks, although two picks can be fit into one pocket. Decide where you want the pocket, keeping in mind that ease of access, bunching or hanging when you sit or stand, and natural body hollows are of primary importance. The best place may well be inside an existing pocket. Other low-profile locations are the small of the back (but lousy access), just inside the lower pant leg, just inside the lower long sleeve, and inside the breast pocket. The iron-on pocket may be showing or actually affixed to the inside of the garment.

After deciding on the size of the pocket (which is easily done by laying out the number of picks you want to carry on the cloth, and drawing outlines around them, also allowing three quarters to an inch for an all-around margin), cut the cloth to size, and wrap the picks in two thicknesses of aluminum foil.

Once wrapped, the picks should be taped in place on the fabric. Try to use double-faced tape so that the tape does not

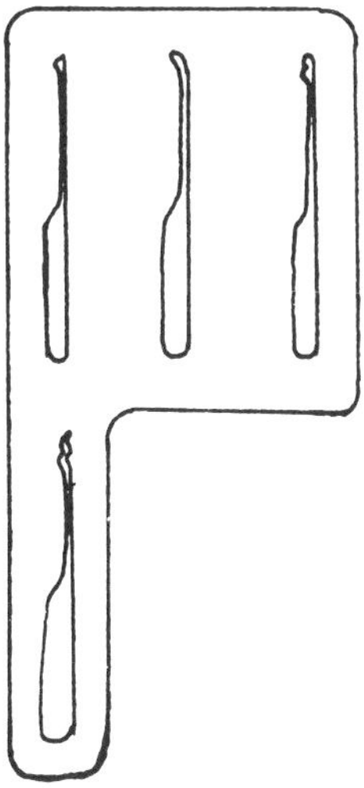

Figure 14.

protrude past the pick border. Wherever the tape is, the pocket will not stick. Once taped in place, the fabric/pick complex should be placed on the garment in the appropriate spot. Need I say that the ironing board should be ready, as well as a hot iron, according to the manufacturer's directions? If desired, the fabric/pick complex can be held in place with some tape, but very sparingly, as the iron may make the tape bleed through the cloth.

Next, iron the fabric firmly, and the thermo-plastic adhesive on the fabric will soak into and adhere to the garment -- be sure to heat well. After this make a test (most directions will also say this) to see if the fabric "took" well. If not, it must be reheated until it does. Now let it cool. Do not attempt further work until it is very cool.

The final stages consist of extracting the pick complex (aluminum foil and picks) from the fabric pockets. If you were smart you put foil all the way up the edge of the fabric patch -- otherwise you sealed the pick forever and now must cut it out. A little judicious cutting may be necessary, anyway, to extract the picks and as much of the foil as possible. When completely foil-free, dust a little talcum powder into the pocket or a little (very little) vaseline to insure ease of pick insertion.

The fabric patch can also be worked or repeatedly bent and rolled until the stiffness it creates does not show, especially at the edges. Of course, the picks will stiffen the pocket when inserted, which is why some locations over the body hollows are better than others.

A removable carrying case is much different, but not necessarily more complicated. Now is the time to take a tour of the local fabric store. You will be looking for several items. First, some heavy or medium weight vinyl fabric, or some of the new synthetics that look like leather. Get a quarter yard at most. Also on the list is velcro, heavy-duty type, in either strips or pre-cut shapes. Also pick up some artificial leather belting material (many different brands and types) or some woven belting (again many different widths and styles), about one-half to one inch in width is good. If you don't like the security of velcro, pick up some heavy-duty closures like buckles or snaps, also.

Now the simplest type of tool pouch you can make from all this is called a "tool roll". Simply lay out your picks and other pocket-size tools in any convenient arrangement, remembering that you can put two or three picks in one location, stacked. Now trace the outlines on the sheet of cardboard they were laid out on, figure an inch margin all the way around, and cut out the cardboard (now it is a pattern). Next, lay the pattern over the fabric, securing it with pins or paper clips if needed, and cut out the pattern. Now for the easy part -- just make a cut with a razor blade, through the pattern at each pick stack, perpendicular to the long axis of the pick, and at the height where the pick goes from handle to shank. Make the cut only as wide as the pick.

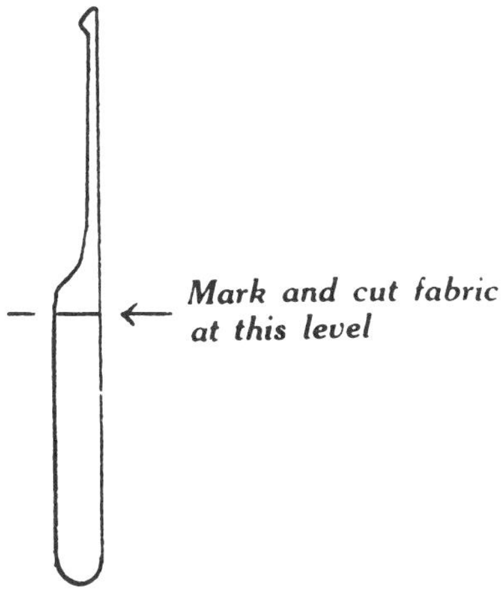

Figure 15.

Now remove the pattern from the cloth, attach it to another piece of cloth just like you were repeating the procedure, but cut out the cloth without the slits. Remember not to flip the pattern over if it is "handed". Keep in mind however, that since the two pieces of cloth must join backside to backside, it will be necessary to flip the cloth over for the second cloth cut.

Now take complete cloth pieces and join them backside to backside, leaving the good sides facing out. The joining can be accomplished a number of ways. The easiest is to use fabric glue, or "instant-sew", as it is called. Test to see if the solvent melts the kind of vinyl or synthetic that you are working with, and if not, start to apply the glue to the backside of the piece with the slits.

As figure 16 shows, apply the glue in a "U" shape around the slits so that when the two pieces are joined, they will form

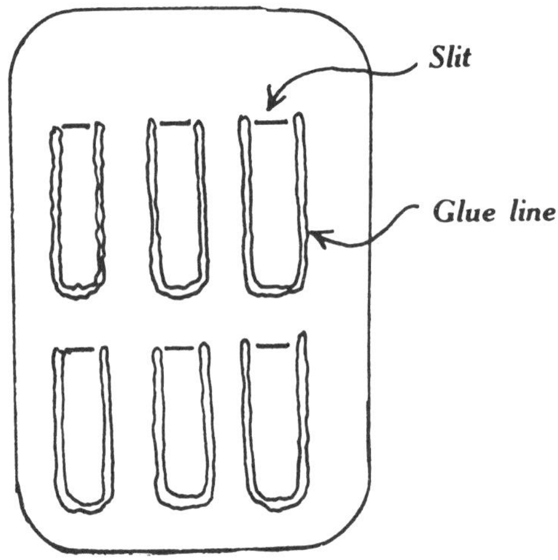

Backside of slit piece

Figure 16.

pockets, and the slits will be the pocket access. Sewing the two pieces together is also possible, particularly if you have access to a sewing machine. It is very easy, or you can con someone into doing it for you.

As a final touch to this construction, install a velcro closure along the two edges of the pouch, either one continuous strip, or a series of pre-cut shapes. Again, either gluing or sewing will work. If you used glue to join the two pieces, you should consider buying a roll of cloth tape an inch wide, and using it to edge the pouch to prevent fraying or separation. Just cut a piece of tape to length and stick half of it on one side of the pouch, then fold over the other side. Burnish this tape down well, and roll it back and forth to make it flexible.

If even this is too much work for you an easier solution is to cut not one but two slits both slightly *narrower* than the pick stack, and the second slit about one quarter of an inch below

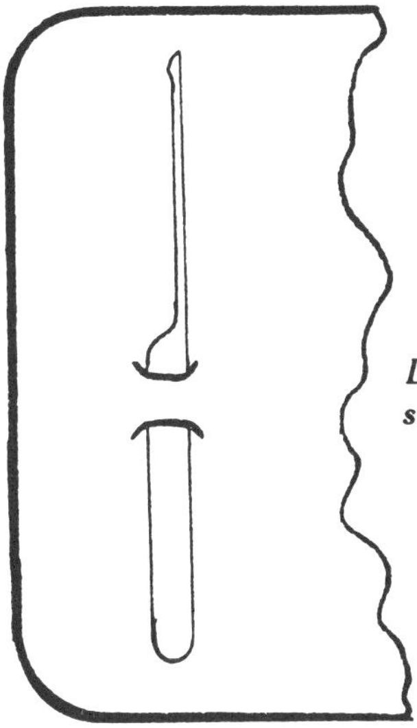

Figure 17.

(towards the handle) the first slit. If you don't cut the slits too long, the pick will fit into the loop created by this cutting, but the fabric must be rolled up and secured with velco, a rubber band, or a cord, or the picks will fall out.

If the machine that you have access to does button holes, consider button-holing all of the slits for a nicer appearance and resistance to tearing.

Another method of construction is to mark the cloth while it is still attached to the pattern with the outlines of all the tools. To do this you need to pierce holes at intervals around the pick outline on the pattern, and then mark through these holes with a marker or ball-point pen when on the fabric. This gives a "dotted-line" impression of where the picks are. Once that is done you can tack (by sewing) an end of fabric strip as wide as the handles of the picks are long, and as long as three times the pattern width to the outside edge of the pick stack nearest the edge of the pattern.

Now insert a pen or lipstick where the pick would go, and tack down the other edge of that first pick stack, making a pocket with the wide strip.

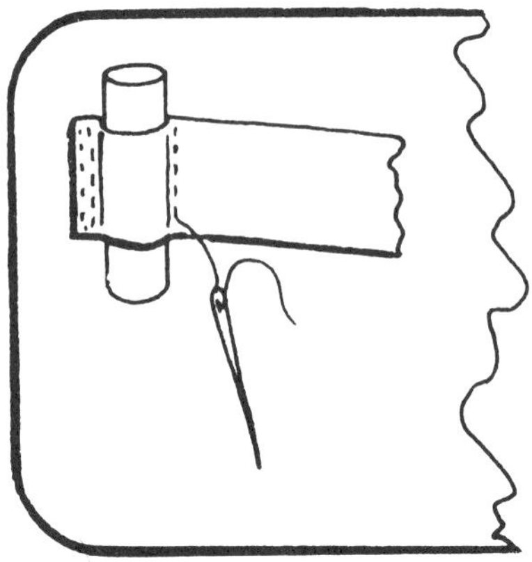

Figure 18.

If you are putting multiple picks in that particular pocket, check at this point to see if they will all fit. You may need to adjust the size of the cylindrical holder. Also, in a pinch you can tack down the strip on both sides of the pick outline by using pins. Once the strip is completely tacked down to all pick outlines, forming a series of pockets (you can cut the strip at the end of a row of pockets so as to continue to the next row) go back and securely sew the strip along the two pick outline edges, and also make a long stitch straight across the bottom of the strip, forming a set of little pouches.

If there are other ranks of pouches, do these at this time also. This tool roll can be velcro-fastened at the edges, or tied as usual.

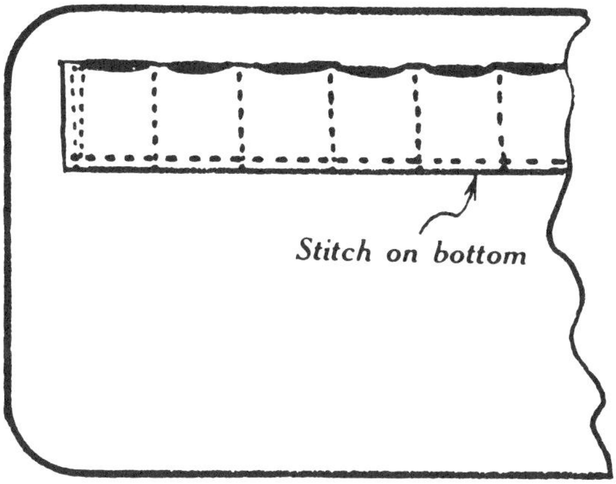

Figure 19.

An interesting variation on this basic pattern is the wrist tool kit. Remember the belting or strapping you got at the fabric store? Cut two pieces to length, one to fit around the wrist, and the other one about six inches up the arm. After cutting, apply snaps, buckles or velcro to the ends.

The next step is to produce a tool roll for one or two stacks of picks, and six inches long. When this is completed it should be then sewed or glued to a piece of flexible plastic, sturdy cardboard, or stiff belting material. The stiff backing will act as a spacer to the two wrist bands and also support the picks. By the way, a piece of stiff plastic can easily be cut from the side of a cheap plastic pail or dishpan.

The next move is to affix the straps to the backing/tool-roll complex, again with glue or sewing. If you want a removable pocket roll, use snaps instead of a permanent fixing.

These are some of the basic principles used to make custom tool holders. I know that an experienced sewing person can easily suggest other more elaborate and more professional-looking pouches. If you are using someone like that, so much the better.

Let me just say a word here about choice of pouch materials. Anything that is porous, like cloth or untanned leather, will not tolerate carrying next to the body. It will rapidly absorb moisture and just as rapidly rust out your picks. Even blued and lubricated picks as per the first section of this book will eventually rust, so pick a non-porous material. If you can work with it, nylon, like the kind that backpacks and tents are made of, if available. It probably will require special glue if you intend to fabric-glue it, though. Of course, any of the vinyls are fine.

Some die-hard agents prefer to carry their picks taped to the skin, so for those let me give a few tips on a body carry. Athletic supply stores carry flesh colored cloth tape in one, two and three inch widths, and they also carry a line of training supplies that should include a spray-on or paint-on skin conditioner and sealer to prevent softening of the skin when repeatedly taped. Also, get a skin lubricant product or some hand cream. When applying a body carry choose an area that is obviously suitable in terms of no excess movement, and also does not perspire excessively. Avoiding hair is also a good idea.

Next, spray or paint the skin conditioner/sealer on and allow it to dry. This will prevent long-term problems. After it is dry, apply a light film of skin lubricant to the pick or picks to be carried, and place them in the middle of the treated area. Finally, apply a long strip of tape to completely cover the tools. Make sure all muscles are fully flexed in that area before applying, or the tape will loosen. An extra tape addition may be necessary in some high-stress areas. It seems like a lot of work, but it will prevent pulled skin and friction blisters. A body carry engineered like this will last comfortably for all day.

I don't recommend any more concealable body carries, except to say that if you are anticipating a skin search, I would secrete a couple of tools in a piece of dense sponge rubber that is jammed into the bullet-shaped plastic tampon inserter so common today. This is strictly at your own risk.

CHAPTER 11
MAKING A TOOL FOR IMPRESSIONING A LOCK CYLINDER

The next discussion is not strictly "tools", but is vitally important. Impressioning a lock cylinder was discussed only briefly in my previous books, so I will treat it in detail here, and show a tool for same. Only impressioning a pin-tumbler cylinder will be covered, because other types of locks can be easily bypassed by other techniques in less time and with less inventory. Inventory is the key word here -- none of the impressioning books tell you that a set of fifty or sixty blanks is required to effectively impression an unknown lock. Some keyways (notably hotels, grand master systems, and government installations) are unavailable in blank form. Nevertheless, let's look into this technique.

When you insert the proper key in a pin-tumbler lock and turn it, what prevents the key from being withdrawn during rotation? A little thought will show you that during insertion, the bottom pins were pushed up past the shearline of the plug and into the top pin wells by the various height key cuts. However, when the plug is rotating, there are no top pin key wells to provide a space for the bottom pins to go if the key was pulled out. So what happens is that as the attempt is made to withdraw the key, the bottom pins are wedged up by the sloping key cuts, and rise until they contact the shear line and the inner surface of the outer shell that the plug or cylinder moves in. When they contact this outer shell, they press against it and stop lifting, thereby preventing further key withdrawal. The item to note here is that they press against the top shell.

Now remember that during the usual picking process, the pin stack can be upside-down picked. That means that the pin stack is totally lifted into the top pin well, and then the stack is raked while the tension is relaxed. At some point, the pin stack will slide down until the shear line of the pin stack meets the shear line of the plug and outer shell. At that point, the bottom of the top pin hangs up on the lip created by attempting to rotate the plug, and that stack is said to be picked.

Impressioning works both ways, taking advantage of both principles. A keyblank is inserted into the keyway and the blank is turned to bind all of the pinstacks. While still applying turning tension, the bow or head of the key is rapped lightly with a hammer, first straight up, then straight down.

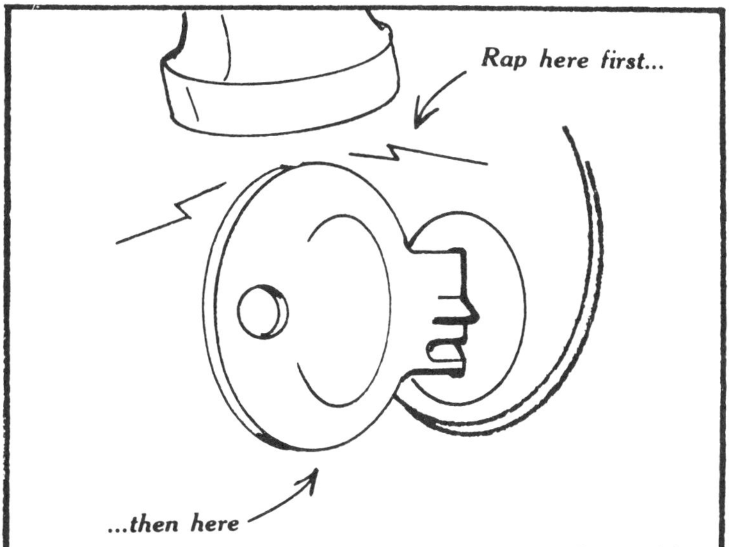

Figure 20.

This rapping produces marks on the keyblank top as the bound pin ends dig into the key surface. Think -- they have nowhere to go when the keyblank surface comes up to meet them, so they make marks in the keyblank.

The next step is to remove the keyblank and examine it for marks. With a half-*round* file, start a keycut at each mark location. Now re-insert the key and repeat the process. A point is reached finally at which one pin stack is lifted up only to the shear line. At that point, when you insert, turn, and tap the keyblank, the top pin is held at shear and only the bottom pin is riding on the blank. This bottom pin alone does not make marks on the keyblank surface, so you know not to file any deeper at that spot. Eventually, all the filing will be done for each pin stack, and the cylinder will turn. Congratulations, you just made a key for that lock!

It's usually not that easy, though, but a little extra knowledge and some practice will make it that easy. First of all, selecting a key blank is important. The proliferation of key blanks on the market of various brands today makes it difficult, but there are not nearly as many keyways, so stocking a set of blanks is possible. The easiest way to get a set is to make a package deal with a small key duplicator, like a dime or drug store. You may even be able to pick up an entire stock and a key duplicating machine to boot. In any event, look for quality keyblanks, try to avoid light metal alloys and aluminum blanks.

Next, be sure that the keyway is well filled, because some master key systems and some keyways vary only slightly in dimension, so avoid a lot of play in the keyblank/keyway fit. A tool that can help you a lot in impressioning is a cast metal turnbuckle fitted with a thumbscrew.

Any well-stocked hardware store carries turnbuckles (not the aluminum extrusion type, but cast) and a machine screw thumbscrew to match the threads of the right-hand side. Be sure not to get a thumbscrew so large in diameter that it will not fit through the smallest hole in the bow of each keyblank you stock. If so, the tool obviously cannot be used. In fact, it may be better to try for two sizes. Also, get two fender washers that are a tight fit on the thumbscrew. The selected keyblank is threaded onto the thumbscrew, using the fender washers if they

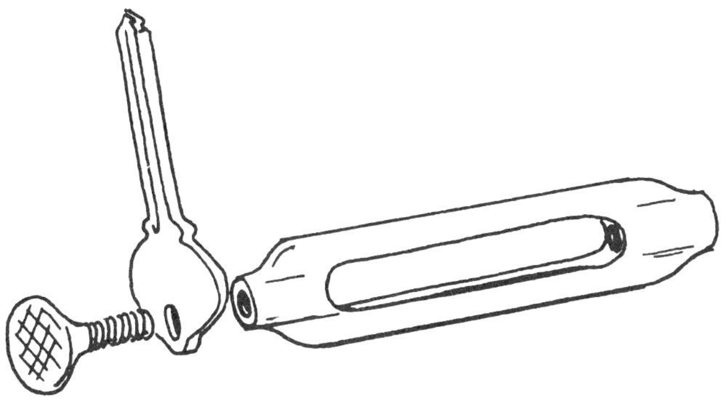

Figure 21.

will not obstruct the keyway insertion, and then firmly screwed into the turnbuckle. The left-hand threaded piece is removed. This combination provides an effective handle to apply turning tension to the keyblank, and also a filing holder. If you are improvising, a metal rod that will fit the hole is also good, but not as firm.

Remember when buying the turnbuckle to pick up a half-round mill file to do the key cuts with. If you wish, you can even buy a vise at some stores that is small enough to carry, or clamp to the top of your toolbox. In a field situation a "C" clamp can be used to hold the keyblank while filing -- or just use any level surface. On to the next step.

Before inserting the keyblank for the first impression, it is usually wise to take a flat mill file and clean off the top .002 of an inch from the keyblank. This is necessary because the manufacturing process produces a hard skin that does not take impressions as well as the softer metal underneath. After this filing (keep the blade top square with the keyblank sides), polish the filed surface with emery cloth to remove scratches. Scratch removal is necessary because the impressions the pin ends make are usually very very faint. In fact, they are not "marks" so much as changes in the glossiness of the key finish. A number of methods all make the marks show up better -- breathing on the key to deposit a thin film of moisture, using a magnifying glass, and the traditional shining a light obliquely onto the key blade. Oblique lighting (like the light of the setting sun) makes small changes in the key surface throw shadows much larger than the marks themselves, so try to master this important technique. See Figure 1. It may be essential if you work in low light conditions.

Once the initial marks have been made (you have to twist hard and also rap hard sometimes to produce a mark), some specialists like to scribe light lines in the sides of the keyblade, perpendicular to the key, that show how the cut is supposed to proceed, namely straight down.

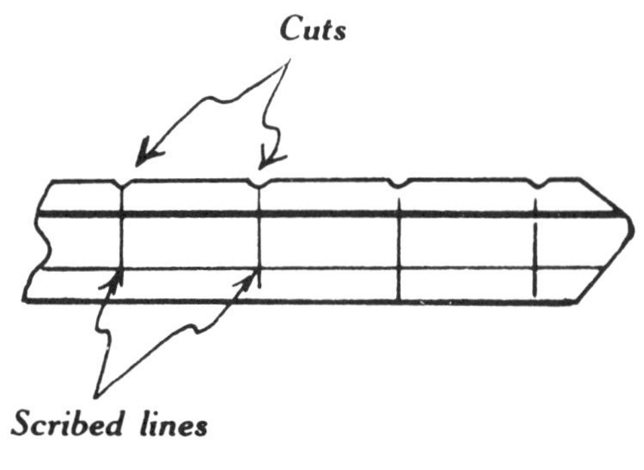

Figure 22.

This has become popular, because if the cut starts to vary (and what hand filing doesn't) from straight down, the pin will mark the side of the cut and not the bottom, making reading impossible. A small machinist's square or a depth gauge will produce the required marks. If you mark the key, remember to follow those marks strictly.

The next tough point is to stop when you don't see a mark anymore. Now the key may not have impressioned well, and the pin stack is not really at shear line yet, but you don't know that, so immediately stop when no more marks are seen. If you get all done (no marks at all) and the key still will not fit the lock, make a little harder impression and look very closely for marks, and one or two may become visible. If you are having trouble seeing the marks, it may be advisable to polish the

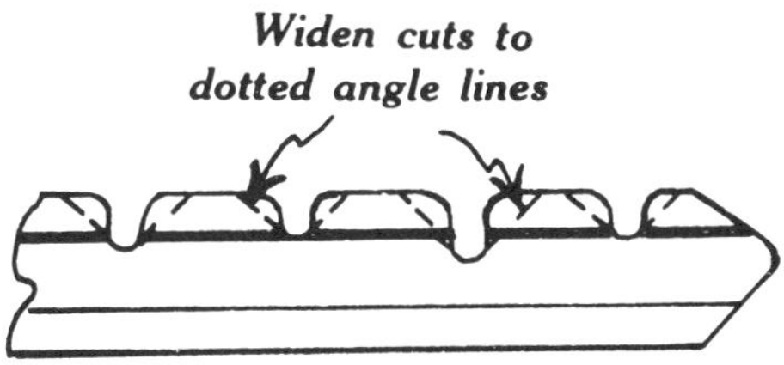

Figure 23.

bottom of the cut with emery cloth each time after filing to produce an easily markable surface. Once the key cuts go to a depth of a sixteenth of an inch or more, it will become necessary to provide angles leading down into the key cut bottoms.

If these are not provided, the pins will catch at the cut bottoms and the key is forever frozen in the lock, requiring disassembly. The angles should be 45 degrees or more -- examine a couple of house keys and duplicate that angle. The actual cutting can be done with a small triangular mill file. Be very careful not to increase the cut depth, though.

Okay, let's assume you just impressioned the key and the lock opened. Touch up and straighten any crooked cuts in the key (but don't alter cut depths), and polish the entire keyblade with emery cloth, especially if guide lines are scribed on the key. Now, as a precaution, blow the keyway out with air to prevent key filings or chips from getting in the pin wells.

Now let's suppose that you cut too deep -- what next? Well, never fear. The easy way to build back the key surface is to carefully rap the side of the blank with a "ball", as in ball peen hammer. This will flatten the metal (you did hit it on the side didn't you?) and effectively make the cut shallower. If the damage cannot be repaired by this method, you must start over.

The final word on impressioning is practice, practice, practice. This skill requires good mark reading ability, but it is the opening skill that most locksmiths use to effect a lock-out or keyless entry, because when completed, the customer has a key that will work. It is also the preferred method of truly professional lock specialists, because once the opening is effected, it can be duplicated in a fraction of the time for subsequent visits. Also, the lock may be impressioned with a rest period between sessions, impossible with other lock bypass techniques, because of the necessity of keeping tension on the lock core.

Just briefly, there is another impressioning system out in

which the key is turned as usual, but then pulled out of the lock about a sixteenth of an inch with great force and slowly (special tool needed for this). This system is not superior, but is easier on the lock. Although disc tumbler locks can be impressioned successfully with up and down impressioning, the pull system will not damage these locks, where up and down usually does. A resourceful person could probably improvise a slammer from a length of rod with a threaded end with two lock nuts on it, a large weight like a roll of washers taped together, and a hook to fit in the keyblank on the other end, but I have not experimented with this idea fully. I prefer the up and down system because of the ease of mark reading. A final caution: use a soft faced hammer when tapping the keyblank or the bow may suffer a lot of damage.

CHAPTER 12
MAKING TOOLS FOR THE SCHLAGE WAFER TUMBLER LOCK

On to the Schlage wafer tumbler lock. There are so many bypass techniques for this lock that I will concentrate on the tool-oriented one only. I strongly suggest that you get one of these locks, and two keyblanks to fit it (there are two different keyway profiles). After disassembling the lock down to the core, insert the key that fits the lock and notice how it pulls some wafers into the core, and *does not* push others out. Notice how a key blank will retract some wafers but also push some out. The object is to get a key that is selective, retracting all the wafers that should retract, without pushing others out (for complete technical information, see my book THE COMPLETE GUIDE TO LOCK PICKING).

After inserting the regular key, again notice where *uncut* portions of the key push out wafers. These must be cut off the tool. To proceed, match up the key and keyblank one on top of the other. Notice the extreme cut at the tip of the key. If the hollowed out side of the key (warding) is away from you, and the cut is on the bottom key half, it is a "type one" -- if on top, it is a "type two".

Again aligning the two, scribe a line at the same depth as that tip cut onto the keyblank, using the cut key as a guide. Now separate the two, and using a ruler, continue that cut straight back, parallel with the key warding right through to the end of the key. Even cut off that part of the bow that lies outside the line. You should come close to the hole in the elbow. Now on the other side of the keyblade, duplicate that depth of cut, except for the tip, which should remain uncut (no cut in the tip wafer position). The cut should end in the usual angle where the key bitting usually ends (the first wafer position).This means that the opposite side of the key will not be cut in the tip wafer position which is called the master wafer, or past where it is cut for the first tumbler

The tool set is completed by getting a keyblank that is the other type (the one that you didn't have) and cutting it according to these same directions to produce two asymetrical blanks, one for "type one", and one for "type two". Remember, the side with the tip cut is totally levelled -- the opposite side is only cut in the wafer positions and not in the master or tip wafer position. Now get a short machine screw or a mill end rivet and fasten the two together by the bow holes.

The other pick needed for this key pick set is either a straight pick or a length of wire. Either must have a sharpened wedge end to get under the wafers. To use, insert the key (either one) and then, using the feeler pick, manipulate all of the tumblers at the side of the keyway accessible because of the flat cut, so that they are all straight. You will be pulling some wafers into the plug, and unfortunately, pushing some back out. A feel or resistance at the end of the pick's travel indicates that the pick is contacting a master wafer. If so, you guessed wrong and you need to start over with the other key. This entire procedure only takes twenty seconds.

Now finally you have the right key, and all of the wafers on the sheared side are lined up. Now the plug is turned (apply tension) and the straight pick is withdrawn. Now the only thing holding the plug from rotating is the tumblers that were pushed out of the plug as the others were drawn in by the straight pick. Slowly, very slowly, decrease the turning tension, and at some point these hanging tumblers will unbind and retract into the plug. At this point, the lock opens. If it doesn't try again. A tool like a pick key leads people to suspect that the picking is almost automatic. The truth is however, that the key is just an easy way to hold large numbers of tumblers -- the manipulation is still difficult and requires practice. You will find that there are two keyway profiles available, and perhaps a third in some localities, so at least two sets of keys should be available. This lock is in widespread use, so bypass tools for it are highly desirable.

CHAPTER 13
CLOSING REMARKS

This concludes the lessons for HOW TO MAKE YOUR OWN PROFESSIONAL LOCK TOOLS, VOLUME TWO. Be sure to see my other books (HOW TO MAKE YOUR OWN PROFESSIONAL LOCK TOOLS; and THE COMPLETE GUIDE TO LOCK PICKING), for the finest in lock information from someone who has experimented and done.